rural routes
and networks

creating and preserving routes that are sustainable,
convenient, tranquil, attractive and safe

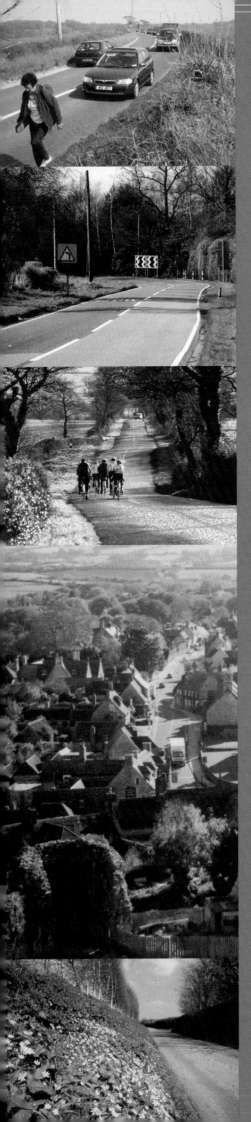

A 1000% increase in motorised transport over the past 50 years – with sustained growth forecast – has put massive pressure on the countryside

Rural routes and networks investigates methods of…

Encouraging the use of sustainable transportation by providing safe joined-up networks for non-motorised users

Protecting and maintaining the local rural character of and around highways and byways

Contents

ISBN 0 7277 3203 X

© Institution of Civil Engineers, 2002

Published for the Institution of Civil Engineers by Thomas Telford Publishing, 1 Heron Quay, London E14 4JD

Designed by Kneath Associates and printed by Latimer Trend

Photo credits: Scottish Environment Protection Agency
GL Jones, Sustrans
Allan Wheeler, Transport Research Laboratory
Mary Weston, British Horse Society
Countryside Agency

Introduction

The key issue

Motorised traffic has grown by around 1000% over the past fifty years and further growth is forecast for decades ahead. Vehicles are faster, safer and cheaper. The roads on which they run have been smoothed and widened to accommodate them.

The changes have brought major benefits but also costs. Increased traffic has made life unpleasant and unsafe for people using the roads, which form a vital part of the network for walking, cycling, horse riding and other forms of movement. Sections of footpath and bridleway have become cut off by busy roads.

Too often rural roads have been marred by utilitarian and insensitive design. Much of the tranquillity and natural beauty of the countryside has been lost to creeping urbanisation and the spread of noise and light pollution. We are all affected by these losses.

It is important to recognise the many different roles the network of roads and public rights of way play in supporting rural life and rural economies. The network needs to be managed as a whole. It is wrong to treat pedestrians and riders as a minority group in a separate budget.

It is also important to realise that small, unnoticed changes from one year to the next, amount to major changes over a decade.

Growth in vehicle use

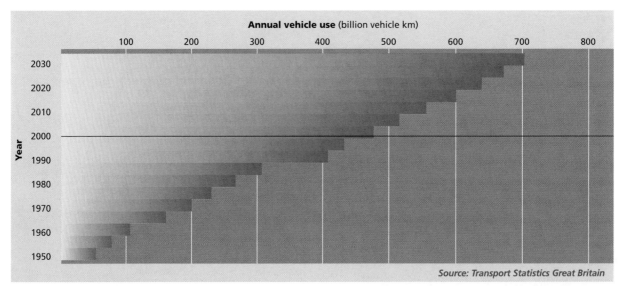

Source: Transport Statistics Great Britain

Choosing and changing the future

We need to recognise the importance of rural routes, the contribution they make to people's quality of life, and their role in the increasingly diverse rural economy.

Change for the worse has largely been accidental. Change will be for the better if professionals, public and politicians work together. The purpose of this report is to stimulate debate and long-term widespread change among:

- **Professionals:** including engineers, rights of way officers, and planners

- **Users:** from motorists and truck drivers through to riders and walkers

- **People:** who live and work in the countryside and those who visit it

- **Politicians:** and community leaders

- **Today's students:** who will be tomorrow's decision-makers and beneficiaries

The report invites people to take a fresh view of how the countryside is changing, to use their judgement, and to ensure that the future we build is one we want, not the one we could have avoided.

A vision and a challenge

- Re-create a network of safe and pleasant rural routes for walking, cycling and horse riding

- Provide for the people who live and work in the countryside

- Restore and protect the character and natural beauty of the countryside

Many solutions and much ingenuity are needed.
We all have a role to play.

Problems

Problem 1: Broken networks

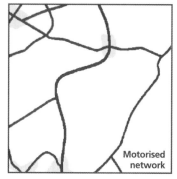

Motorised network

Walking network

Cycling network

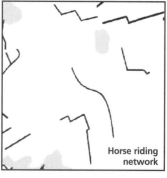

Horse riding network

The footpath and bridleway network evolved in an age when most travellers and workers walked, and only a few rode horses or travelled by carriage. Continuous and convenient routes were provided for all users.

In the first half of the 20th century, motor traffic on rural routes was so light as to have little effect on other road users. But the increase in traffic over the past 50 years has turned many roads into unpleasant and dangerous places for walkers and riders.

The accompanying diagrams illustrate what happens. The motorised network is well defined, joined up and easy to use. But walkers and riders find themselves using a network broken by sections of busy road where motorised transport dominates. It is great fun walking, or riding, but it is no fun if parts of the journey involve confronting fast, heavy traffic.

Past approaches by local authorities to join up non-motorised networks have been based on one-off projects. If walking and cycling in the countryside is to become an attractive and safe alternative to travelling by car, joining up networks needs to be a core policy objective for local authorities.

Safety

Overall, the number of casualties on our roads is declining. But over 60% of fatal accidents on A roads occur in rural areas.

The speed of traffic is substantially higher on rural roads than on urban roads.

The severity of injuries in a crash is directly related to the speed of the traffic.

Local authorities have a duty to improve road safety. If joined-up networks were available for pedestrians and riders, they would not often need to use busy roads. Safety would improve.

Break-up of the walking network

Most journeys on foot involve some travel along a road. The growth in the volume, speed and weight of traffic makes walking along roads unpleasant and unsafe. Sections of footpath can become isolated.

The problems can be at their most acute with footpaths that lie on the edge of urban areas. These paths are among the most valuable and heavily used parts of the footpath network, but often their use involves walking along sections of very busy road.

Despite the large number of people able to walk, increasing numbers find it safer, more convenient and more pleasant to take their cars.

65% of users feel threatened by traffic "all or some of the time"

91% think that the speed limit should be reduced

72% think that vulnerable users should be given priority on selected country lanes

Source: CPRE "Rural Traffic Fear Survey", 1999

Of our population of 60 million, around 5 million occasionally or frequently walk cross-country for pleasure.

How many people would walk but do not because of the danger?

3.1 million people cycle every week

Since the mid 1980s the number of journeys made by bike has fallen by 36%

Source: Transport Statistics Great Britain

Break-up of the cycling network

Roads and lanes make up a substantial proportion of the cycling network. A typical cycle journey will involve negotiating potentially hazardous roads.

The fatal accident rate for cyclists on rural roads is four times that for urban roads. Cyclists are permitted to use bridleways but few have surfaces smooth enough for cycling.

How many people would cycle but do not because of the danger?

Over a million people ride horses in the course of a year, but increasingly the use is off-road

100 horses are killed on our roads every year

Source: British Horse Society

Break-up of the horse-riding network

The bridleway network is limited, and roads and lanes make up a substantial proportion of the total horse-riding network.

How many people would ride but do not because of the danger?

Problem 2:
Urbanisation

A rural road contains just three elements: the road itself, the countryside and the sky.

This simplicity supports a tourist industry worth billions of pounds per annum, but across the UK we are witnessing the gradual erosion of this simple character. Some roads are unaffected; others change beyond recognition over a 20-year period. Much of this change is thoughtless and unnecessary.

Changing roads

Non-native hedge

6ft fence

Suburban gates

Grand entrance

Intrusive lighting

White lines

Concrete kerb

Visibility splay

Poor development

Advertising

Fly tipping

Dumped cars

Eroded verge

Rural roads are a key part of the rural landscape. From lowland Devon's narrow lanes lined with Devon banks, to the winding hedge-bordered lanes of the weald, the open moorland roads of Scotland and the drove roads of central England, there is so much rich variety. The appearance reflects the local geology and plant life, local craft traditions, and the function that the road has performed over the centuries. However, rural roads can easily be turned into sterile, uniform corridors that carry an industrial landscape into the heart of the countryside.

Further urbanisation occurs by changes to the land next to the road. Hedges are replaced by 6ft fences, property entrances are given the grand treatment with wrought-iron gates, brick pillars and wide visibility splays, gardens are extended and non-native trees introduced, so that rather than the land and buildings blending into the countryside, they are little parts of the town that contrast starkly with the rural landscape.

The precious character of each part of the countryside is slowly vanishing.

Time for a rethink

Rethinking... existing practices

To preserve the rural landscape and ensure that there are quality routes for walkers and riders, we need to look again at the practices used in route management, planning and construction.

1. **Review the balance of funding** allocated for motorised and non-motorised routes. Set aside a certain proportion of the motorised route budget to redress past under-funding in the non-motorised route network. The percentage could start relatively low (e.g. 10%), rising 2% annually. Rights of way departments have traditionally had nominal budgets. A small funding increase would make a significant difference to the non-motorised route network.

2. **Set priorities.** Consider the adoption of a formal road user hierarchy to help determine priorities for funding and action on different routes.

3. **Set clear timescales and goals** for correcting accumulated damage to the walking and riding network.

4. **Manage demand** in society's overall interest. Avoid a "predict and provide" approach to accommodating ever-increasing numbers of vehicles. For example, use measures to make driving less attractive, provide high-quality public transport and increase provision for walkers and riders.

5. **Join up the networks.** Aim to create functional and attractive route networks for all users. Ensure integration of transportation, highway and rights of way strategies, rights of way improvement plans, walking, cycling and equestrian strategies; and compatibility with local transport plans.

6. **Provide safe and attractive conditions for shared use:** e.g. reduced speeds, Quiet Lanes, user segregation, or provision of new non-motorised routes based on potential demand.

7. **Introduce a "Net Gain Principle":** i.e. that any newly built or improved route for motorised traffic also benefits the non-motorised route network.

8. **Practise flexible approaches to design.** Practitioners should be encouraged to think about causes of problems and innovate, rather than unthinkingly applying standard designs.

   ■ A clear distinction needs to be made between mandatory and advisory guidance documents. There is much more latitude than is commonly believed.

■ Does everything have to be signed or can the concept of "self-explaining roads" be adopted?

■ Can we develop "rural design" for rural roads – rather than translate urban highway engineering into the countryside? The standard *Design Manual for Roads and Bridges* is for major roads – it does not have to be followed for country lanes.

9. **Highway engineers and rights of way officers should work in partnership** with local communities and user groups, as well as with landscape architects, conservationists, etc.

10. **Use all available resources.** Involve the community and users through means such as Local Access Forums. For finance, options range from transport budgets through to regeneration and tourism.

11. **Share best practice.** Advertise your successes and experiments to a wide audience. Keep abreast of the latest methods for travel management in rural areas, both in the UK and elsewhere.

12. **Develop technology.** Pre-empt and support the introduction of new forms of traffic management, such as Intelligent Transport Systems.

Learning from other countries

There is a need to look at and begin to widely apply the best forward-looking practice from overseas, especially Europe. Germany led the way in user segregation by being the first country to make widespread use of motorways (*Autobahnen*) around 70 years ago. Most developed countries have now followed this lead, but it took the UK until the 1960s to make substantial progress. For about 20 years now, Germany has also been paralleling most major rural routes with separate walking and cycling paths. They are not alone in Europe. Similar practices are being undertaken in Scandinavian countries and also in the Netherlands, where additional work is being carried out on integrating roads into local landscapes.

Rethinking... choices for roads

It is important to develop a solution that best fits the local circumstances. Many alternative strategies are possible. The following are some simplified possibilities and consequences. They are not mutually exclusive, and can be combined, modified and developed further.

Do nothing

- Motorised traffic volumes will increase
- Accidents could remain stable at best, or possibly increase
- Journeys on foot, by bike or horse will continue to decrease
- Roads will continue to lose their rural character
- Visual, noise and air pollution will increase

Average distance travelled per person per year by mode of travel

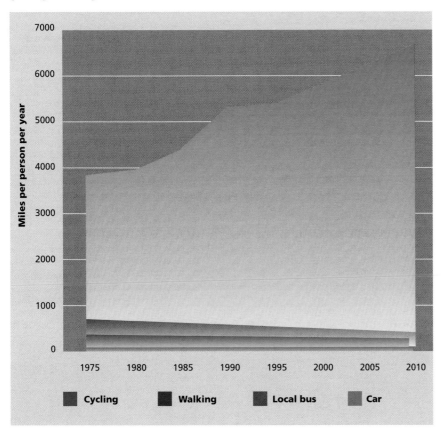

Cycling **Walking** **Local bus** **Car**

Personal travel trends.
- car use... up
- cycling, walking, bus use... down

Source: Transport Statistics Great Britain with projection to 2010

Manage demand

Encourage sustainable demand for movement, alternative modes of transport or routes.

- Assess how people would like the routes to be and try to manage traffic to those levels
- Avoid letting land-use and travel patterns develop that involve use of country lanes by fast traffic. Beware incremental changes
- Try keeping vehicle speeds safe for other users
- Realise that small localised highways works may encourage more traffic. For example creating additional or wider passing bays, widening of lanes, easing of bends.
- Use awareness campaigns or advertisements to encourage people to use different forms of transport on the existing network
- Publicise a well-thought-out route hierarchy to enable users to better understand how to use the existing route network
- Restrict motorised access on certain routes to encourage people to travel by other means
- Consider changing the *Highway Code* to give non-motorised users clear priority on appropriate routes

Manage for combined use

Manage the road for combined use by incentives, reducing speeds, introducing traffic calming or establishing shared surface schemes, e.g. Quiet Lanes or Greenways (see next section).

- Introduce measures to discourage unnecessary traffic, e.g. providing restricted entry widths
- Keep signs sensibly sized, to cater for all users, and to be sensitive to the setting
- Consider widening verges for walkers and riders or providing side strips (a section of the road denoted by a different surfacing material)
- Consider the concept of an absence of road markings denoting a "low-speed country road"
- Engineer for slow speeds
- Match the traffic to the roads, not the roads to the traffic

Develop separate routes

Site footpaths, bridleways or cycle tracks well back from the existing road (e.g. behind a hedge, wall or bank), or designate an entirely new non-motorised route following "desire lines".

- Ensure that motorised traffic is kept slow and safe on non-motorised linking routes
- Create good-quality cycle lanes or a segregated shared cycle track next to the road
- Ensure that crossings and bridges are integrated with the walking and riding networks
- Only widen a lane if it is to be upgraded to a through route with off-road facilities for walkers and riders

Rethinking… networks

Roads form an important part of the network of routes for walkers and riders. They need to be managed carefully if their value is to be preserved.

Horse riding and shared networks

Make it possible for horse riders to share roads safely with walkers, cyclists and motor vehicles, by reducing traffic speeds or introducing appropriate traffic calming.

Where possible, but especially on busy roads, ensure that grass verges are preserved for horse riders. These are an essential element of the horse-riding network.

Quiet Lanes, such as this one in Norfolk, aim to encourage mixed use.

Cycling networks

Cycling offers a direct alternative to travelling by car for short journeys. Provision of convenient and safe routes is the key to getting more people to cycle. In recent years, many new motor-traffic-free routes have been developed within rural areas (many forming part of the National Cycle Network – NCN), with surfaces suitable for all types of bike. These can form important elements of the cycling network, and need to be integrated with on-road routes.

As with walkers and horse riders, busy road crossings can be a major barrier. Adequate facilities need to be provided at such locations, and often a short length of traffic-free path alongside the road will be required to reach the next on-road section.

- ■ Care needs to be taken in resolving conflicts of use with walkers and horse riders, especially where there is shared use.
- ■ Secure and convenient cycle parking should be provided at key locations and attractions.
- ■ Much of the NCN has been created by signed use of minor roads rather than new construction.
- ■ Canal towpaths can form a useful part of the network.

Walking networks

Busy roads blight life for pedestrians just as much as for horseriders or cyclists. What can be done?

Providing a suitable path

The type of path and route provided should match the users.

People walking to work, to shops, to school, or perhaps to other public transport, will generally require a surfaced route. Lighting may be required if people are to use the route after dark, especially if personal security is a local concern.

Routes for recreational use can often be unsurfaced. Where the users are out for recreation, and equipped for mud, then it should be straightforward to provide alternative routes to avoid busy roads.

Footpaths often provide more direct routes than main roads. They also eliminate danger from passing traffic and are more pleasant to use.

Improving conditions along busy roads

The standard solution to improving conditions for pedestrians along a busy road is to construct a footway immediately next to the road. It is often the easiest option, but there are two reasons for thinking twice about this:

1.   Urbanisation – A roadside footway widens the tarmac, and can destroy verges, hedgerows and the road's rural character.

2.   Safety – Roadside footways place pedestrians next to traffic that may be travelling at 60mph or more. Passing traffic can offer some surveillance, but the traffic also brings noise, fumes and danger: 8% of pedestrian casualties occur while the pedestrian is actually on the apparent safety of the footway.

Pedestrian safety and enjoyment are improved by placing the path away from the road.

Consider alternatives before a decision is made - including:

■   reducing the speed and/or volume of traffic
■   creating a new or improved footway by reducing the width of the road – and retaining the verge
■   providing a pedestrian route set back from the road separated by a grass verge, wall or hedge.
■   creating an entirely new pedestrian route that is on a more convenient or direct route.

This footpath has been set back from the roadway. It is more pleasant to use than a footway and is undeniably more attractive.

Rethinking… highway engineering

The appearance of the road plays a major role in our perception of the countryside. Rural highway engineering should reflect the character of the local countryside, including landscaping and conservation. Safety and accessibility for all users should be fundamental to how a road is engineered.

Junctions

Many accidents occur at or near junctions. Some of the changes that have been introduced to improve use for motorists have had a detrimental effect on others: swept curves at junctions increase the speed of traffic and widen the road at pedestrians' crossing points; wide, deep-visibility splays change the character of an area by introducing areas of mown grass and removing hedges and mature trees.

To maintain a rural appearance and provide safety for all users, try to keep junctions tight. If access for large vehicles is an issue, provide an overrun area in preference to a wider junction. Reduce traffic speeds in preference to creating wide visibility splays. Be flexible over how the visibility splay is managed: try to make the appearance blend with the surrounding countryside.

This splay in a national park has urbanised the junction by introducing an area of mown grass.

At this junction, the splay is narrow and signing is understated.

Surfacing materials

When travelling along a road, the road surface forms about one quarter of the view. The road surface can add to the attractiveness of a route or a village, and reinforce local character. It can also be used to signify a change in priority, e.g. to suggest a shared surface. A wide range of different surfacing materials and colours is now available as alternatives to standard blacktop.

In villages try to follow local traditions and use locally sourced materials where practicable. Nationally

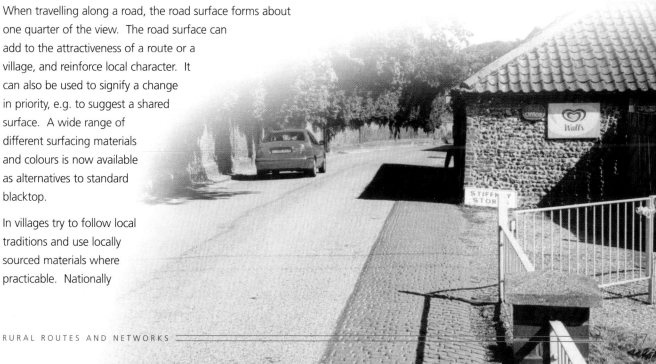

sourced "heritage" materials, for example granite setts or clay paviours, will not necessarily be the right choice for country villages where rammed earth and stone was traditionally the most common road surface. A simple bitumen surface may actually be closer to the original.

Recycled materials are sometimes used to improve the utility of footpaths and bridleways. Be sure that careless use of these materials does not destroy the attractiveness of the right of way.

Select surfaces suitable for the users. Consider horses, cyclists, pedestrians, wheelchair users and vehicle users.

Kerbs

Kerbs urbanise lanes, and by reinforcing the line of the road can encourage speed. By changing the drainage they may cause other problems elsewhere by concentrating run-off which may then involve the expense of installing pipes and gulleys. Avoid them if possible.

If vehicles are over-running on to verges, consider whether it is really a problem that needs solving, or whether it is self-limiting.

An alternative to laying kerbs could include planting or building up the verge.

If some sort of kerb is necessary, try to use low-profile bevelled kerbs or preformed channels. If the problem is one of drainage, try to use swales (wide ditches designed to store and slow the flow of water) or an alternative sustainable drainage solution.

Drainage

Standard drainage systems urbanise the countryside. They may also require further construction works down stream. Natural drainage methods help to reduce run-off, provide some protection from pollution, and provide better value to wildlife. Concrete pipes or culverts tend to get blocked, so it is better to avoid culverting ditches or watercourses. Gravel-filled ditches are visually intrusive – a simple ditch or a grass swale will match the rural landscape better.

Left: This swale avoids the need for a piped drainage system altogether. Rainwater runs off the road, over the verge and into the swale. Pollutants are trapped in the swale, and the flow of water is slowed, reducing the risk of flooding downstream.

Middle: An ugly, standard design: alternatives are available. Had sustainable drainage systems been used, much of the construction expense could have been avoided.

Right: A standard design feature – urban in appearance.

Top left: before and after shots of a road, after the removal of white lines, the introduction of a 20mph limit and the laying of an uncoated stone surface.

Left: Signs, lines and coloured high-grip surfacing warn drivers of an impending corner, but with significant visual impact.

Road markings

On main roads, road markings can be important in aiding the safe movement of traffic; in some areas, they are unnecessary. Try to keep markings to a sensible minimum. If possible, avoid them altogether on lanes. Should we be using the idea of self-explaining roads more widely?

Traffic calming

Traffic calming in the countryside is often urban traffic calming with light modification. It can be difficult to achieve reductions in speed without intrusive measures. Options include narrowing roads, introducing slight bends in the road, and using trees and landscaping to try to make the appearance of the lane encourage slower speeds.

It can be difficult to achieve substantial reductions in speed without intrusive measures.

New roads

Consider the Countryside Agency guidelines (*Roads in the Countryside*: CCP 459) when constructing new roads.

Try to create a new road that will become a seamless part of the surrounding countryside and local road network.

Ensure that local building styles and landscape characteristics are maintained, and that no land is "wasted".

Think about walkers and riders: in some cases it may be favourable to provide an adjacent path away from the road.

This new 60mph road did not follow the Countryside Agency's guidance:
- **The fence follows the skyline, making it very visible**
- **The positioning of a footway on the outside of a bend puts pedestrians at direct risk from out-of-control vehicles**
- **More gentle grading of the cutting would have allowed the land to be returned to agriculture.**

Noise versus tranquillity

Peace and tranquillity are hallmarks of the countryside. But the hum of insects, birdsong, and the gentle rustling of leaves in the trees can be drowned out by traffic noise. Roadside noise can reach 90 decibels, can disrupt birdsong for several hundred metres, and can carry for several miles. Tyre noise is the main source of noise when vehicle speeds exceed 20–30mph.

Over short distances sound travels on a fairly direct path; over longer distances the sound we hear tends to come from out of the sky – curved down by wind or temperature gradients. So noise can be reduced over short distances by barriers, but over longer distances it requires reduction at source.

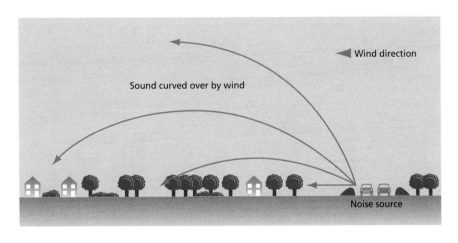

1960's

KMS 0___50

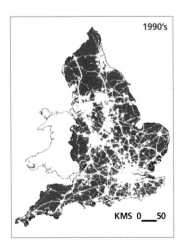

1990's

KMS 0___50

Reducing noise at source

Low-noise surfacing: Low-noise road surfacing materials can reduce tyre noise by around 5 decibels or even 10 decibels over standard concrete roads

Lower speed: reducing speeds by 20mph can reduce noise by around 5 decibels

Barriers

Roadside noise barriers reduce noise over short distances. To have a significant effect the barrier needs to be solid and high – several metres. Noise barriers are available in various materials including wood, metal and reclaimed materials. Bunds or earth embankments can be used to block the path of noise, but to make these look an integral part of the countryside requires skilful landscaping.

Residents should be aware that 6ft close-boarded fences have a negligible effect on reducing noise, and can damage the look of the countryside. Trees and vegetation have some effect on reducing noise when used in depth.

Source: *CPRE*

The CPRE's tranquil areas map shows how much the area of countryside that is largely free from noise or visual disturbance from development or traffic is decreasing.

Rethinking... street furniture

Additional engineering features, such as barriers, lights and signs, should be kept to a minimum in rural areas, and should never dominate. Think carefully about how street furniture can make the route safer, more accessible and better looking for *all* users, not just motor traffic.

Safety barriers

Wire rope barriers are less intrusive than sheet steel but require a long run and depth to allow deflection. Use native screen planting that conforms to the local landscape to hide sheet steel barriers. Place barriers between traffic and people.

This barrier protects the bridge, but not the people – as evidenced by the tyre marks on the pavement. The barrier could have provided protection for pedestrians as well.

Lighting

Consider how the lighting will appear both at night and in the daytime. Keep light pollution to a minimum by using low columns with flat glass luminaires (full cut-off lanterns). These maximise energy efficiency and minimise visual intrusion, while ensuring that the night sky stays dark. The latest range of high-pressure sodium lamps can provide a near white colour and should be considered. For further information see *Lighting in the Countryside – Towards Good Practice* (www.planning.odpm.gov.uk/litc/index.htm).

Left: Light pollution in this Area of Outstanding Natural Beauty is caused by poorly designed lights.

Right: Modern flat-glass luminaire: high-efficiency illumination and minimal stray light. Older design: considerable stray light contributing to sky-glow and wasting energy.

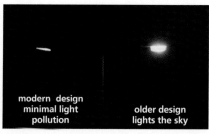

modern design
minimal light
pollution

older design
lights the sky

Left: Use of traditional road signs in this area contributes to local distinctiveness.

Top right: Advertising can become widespread on busier roads. Planning powers can be used to keep it under control.

Bottom right: Which would be the best option in this open landscape – speed roundels, or signs on posts?

Signs

Try to minimise signage and advertising. Keep it discrete.

In some areas there are local styles of finger post signs in use; including wood, cast iron, white on black, black on white.

Think about improving direction signs for off-road networks, to encourage more people to use existing footpaths and bridleways.

Other street furniture

Keep street furniture to a minimum. Eliminate unnecessary posts. Design street furniture to fit in with the surroundings. Mount signs on walls where possible, or in groups. Choose signs of appropriate size and keep them as small as possible; consider whether a sign is completely necessary.

Left: This yellow grit bin in a National Park does not fit in with the local environment. Another colour could have been used to make the grit bin stand out less.

Right: Milestones are part of the tradition of the countryside, and are interesting and useful features for walkers and riders.

Rethinking... use of fences, walls, trees, hedges and verges

Ideas for local authorities and landowners

Boundaries can completely change the appearance of a road. Always consider the landscape context when deciding what type of boundary to use. Rural boundaries are mostly characterised by "soft edges", but in some areas of the UK a stone wall might be more appropriate.

Fences

This fence is an urban feature, and contrasts badly with the natural verge further along the road.

High close-boarded fences destroy rural character. Existing planning controls should be used to prevent the introduction of close-boarded fences over one metre high along lanes.

Follow local traditions, and avoid using close-boarded fencing in rural areas. If a high fence is necessary, set the fence back from the road and plant a native hedge in front to provide screening. Chain-link fencing with concrete posts is characteristic of industrial areas and is best avoided.

Brick walls

Avoid introducing high brick walls in rural areas.

If a brick wall is required, the key advice is to match the materials and practice used to traditional local construction techniques. Follow local practice with brick colour, mortar colour, width of pointing, and bonding (e.g. Flemish as opposed to Stretcher bond).

Check with the planning office or local history societies if you are unsure about which bricks or construction techniques to use.

This wall pays scant attention to local styles and suburbanises the lane. A native hedgerow in its place would match the local landscape.

Stone walls

There is a rich and varied heritage of wall construction in the UK. Try to use locally sourced stone or a close match, and employ local styles of construction and masonry. There are several different traditions of dry walling – be sure to use the right one for your area.

Dry stone wall – Swale Dale.

Trees, hedges and verges

If a barrier is needed, a hedge, densely planted earth mound, or grass verge is often the best option. Always consider the long-term maintenance implications, and use locally found plant species. Use nearby hedges as a guide to local styles and varieties, or consult the Natural History Museum's Postcode Plants Database (www.nhm.ac.uk/science/projects/fff/).

Dry stone wall – Peak District.

Trees make a major contribution to the quality of the landscape and the rural appearance of a highway. Although tree-lined roads reduce drivers' sightlines, they can also make drivers slow down, contributing positively to road safety.

With grass verges, practice management regimes that promote biodiversity by seeking the advice of a local ecologist. Verges can also provide a route for walkers and horse riders.

Hedgerows change with the seasons and provide a rich experience for anyone travelling along country lanes.

Rethinking… land and buildings

Ideas for local authorities and landowners

Gateways and entrances

An entrance to a property can be so subtle that it has no effect on the appearance of a country lane. But some entrances are given an ostentatious treatment, with brick pillars, concrete ornaments and wrought-iron gates. These are urban features.

Gardens

Gardens can allow a building to blend with the surrounding countryside. But thoughtless design can introduce features that jar with the surrounding countryside. Think carefully about the types of trees and shrubs used, and garden buildings. Some tall-growing conifers such as *Cypressus leylandii* have a shape that will stand out in a rural landscape – and will mark the land out as garden rather than countryside.

Buildings

Local materials, styles and construction methods can be used to create new buildings that reinforce local identity and enhance the countryside.

Parish plans and village design statements can be used to encourage better forms of development.

New developments should be designed to be safe and attractive for pedestrians and riders, and link with existing routes.

Follow the latest design guidance, e.g. *By Design: Places, Streets and Movement*, and the Countryside Agency's *Landscape Character Guidance*. For layout information see *Places, Streets and Movement, Home Zone Design Guidelines* – IHIE.

Bottom left: **An example of poor development in the open countryside: the buildings, and surrounding fence, are clearly at odds with the landscape.**

Bottom centre and right: **Is it possible to blend a filling station into a village?**

Taking action

This section aims to help local users to identify the key issues or problems with their current networks. Local users have a unique insight into the standard of network provision, and are well positioned to bring about real change. Communities can apply for funding from the Government (e.g. through initiatives provided by the Countryside Agency), to improve an area.

Three steps for bringing about change

1. Bring people together. Get the local community talking about what they want to happen.

- Bring people together to "own" the problems and opportunities, to provide the basis for action. Use structures already in place – e.g. Local Access Forums, parish and town councils, local user groups, MPs, councillors and residents' associations – to get people involved.

2. Assess the route in question. Get people to determine how a route or place can be improved.

- A "Placecheck" is a series of questions that helps people to agree on how a route or place can be improved. The process is explained further on in this section; detailed questions are suggested in Appendix II.
- Please add the details and relevant problems and issues with your assessment to the discussion forum on the Countryside Agency Greenways and Quiet Lanes website (www.greenways.gov.uk).

3. Community action. Use all available resources to make things happen in your local area.

People:

- Local maintenance/management – work with organisations like BTCV to restore dry stone walls, maintain hedges and manage grass verges.
- Promote responsible use of your local networks – create leaflets to influence local people and politicians.

Funding:

- Get support or sponsorship from local businesses.
- Work with local authorities, community groups and agencies to create or get funding. Options include Vital Villages Schemes, the Rural Transport Partnership (RTP) and Local Transport Plans (for specific projects and programmes for improving routes). Initiatives such as village design statements can help support applications for funding. See Appendix IV for more information on useful contacts.

Placecheck

Placecheck is a method of assessing the qualities of a place, showing what improvements are needed, and focusing people on working together to achieve them.

A Placecheck can start small: with half a dozen people round a kitchen table, or a small group meeting on a street corner. A Placecheck can cover a street (or part of one), a neighbourhood, a village centre, or a lane or footpath. The initiative can come from anyone, in any organisation or sector. It can be used to help produce a village design statement or contribute to a rights of way improvement plan.

Step 1: Look at the network of routes

- Each path, bridleway, lane or road is part of a wider network of routes
- Think about the network of routes, how they are used, and by whom
- Choose one route within that network to do a Placecheck

Examples of different routes:

- Tourist routes which bring trade into the community
- Routes used by farm vehicles and delivery lorries
- Commuter routes to nearby towns or railway stations
- Routes to schools, pubs, local shops and churches
- Routes for recreation/leisure purposes; e.g. walking the dog, evening walks, long-distance rambles, bike rides and horse riding

Step 2: Placecheck
Work through the questions in the Placecheck (see Appendix II).

Step 3: Keep the pace
Implement the findings of the Placecheck.

When the process is complete, it should help the local community to decide what changes need to take place to make the route a better place for all its users – and who can enact those changes.

Quiet Lanes and Greenways

The Countryside Agency's "Quiet Lanes" programme allows the community to take the initiative to solve the problems of their rural route networks. By taking this "bottom-up" approach, the Quiet Lanes initiative can be tailored to suit the needs of different communities. The scheme has been piloted in Kent and Norfolk, with some encouraging interim results. Useful information on how communities can encourage the development of Quiet Lanes in their local areas is available from the CPRE, see Appendix IV for contact details. Technical details and information can also be found on the Countryside Agency's Quiet Lanes and Greenways website.

Greenways are a network of largely car-free routes connecting people to facilities and open spaces in and around towns and cities and to the countryside. They are intended for shared use by people of all abilities on foot, bike or horseback, for car-free commuting, play or leisure. The development process and physical quality of Greenways must satisfy certain criteria: the PACE system is one framework that can be used for the process of developing Greenways and Quiet Lanes. See Appendix III for more details on PACE.

Rural Transport Partnerships and the Vital Village scheme

Rural Transport Partnerships (RTPs)

The RTP scheme is intended to support community initiatives that serve the accessibility needs of residents in any part of rural England. Projects that the scheme can fund include:

Rail, road, water or other transport projects that provide:

- A new or enhanced transport service or facility; or
- Better use of existing resources
- Experimental services (i.e. a proposal combining a feasibility study and action research)

Studies or research proposals (usually part of a wider project) such as:

- Studies of existing service performance
- Transport needs analysis
- Other transport studies

Non-motorised transport projects such as cycle, horse, water, pedestrian or pedestrian-vehicle initiatives, including those which integrate and create links between different modes of transport and access routes or destinations where people live or are visiting.

Further details about this scheme and the finance available can be found on the Countryside Agency's website (www.countryside.gov.uk) under the Transport section.

The Vital Village scheme

Parish Plans

Parish Plans is an initiative of the Countryside Agency's Vital Villages team. It aims to give local communities the opportunity to carry out a parish or town plan. These plans not only embrace assessments like village design statements, but go on to set out a more comprehensive vision for the future of the community and develop an action plan for implementing the vision.

The plan may cover all aspects of future development and may include the following:

- Land and buildings
- Environmental issues
- Transport issues
- Delivery of services and amenities

Financial assistance is available from the Countryside Agency for developing Parish Plans.

Parish Transport Grant (PTG)

This scheme runs until 31 March 2004 and includes projects such as scooter vouchers, car sharing and walking and cycling schemes. A Parish or Town Council can apply for a grant of up to £10,000 (75%) towards the total cost of a transport project. N.B. Only Parish Councils can apply for a Parish Plan Grant.

For more details, or to apply for either grant, telephone the Vital Villages Call Centre on 0870 333 6170.

Summary

The countryside today: successes and failures

Motorised traffic has grown by around 1000% over the past 50 years, but there have been major deleterious and unpredicted side effects. Walkers and riders are being driven off rural roads, either by the volume of motor traffic or by fragmentation of the network. Rural character and tranquillity have been lost, while noise, light, air and visual pollution have all increased. Traffic volumes are forecast to keep growing – so present problems are set to get worse. New thinking to reverse these side effects is needed from built-environment professionals, lawyers, users, politicians and the present student generation, who need to avoid inheriting yesterday's perceptions and whose task it is to develop tomorrow's solutions.

The purpose of this report is to show how and why problems have occurred, so that we can learn from our mistakes and work towards a more sustainable rural route network. This will require new approaches from practitioners and policymakers, and a step change in the way people use the route network. The report aims to start a debate on these issues and institute a long-term process of change.

Our challenges

To reverse the negative effects of increased speed and volume of motorised traffic:

- Community severance
- Increased danger for walkers, cyclists, horse riders, farm animals and other vehicles
- Increased noise, light and environmental pollution
- Reduced use of walking and riding as a means of travelling between destinations

To develop joined-up non-motorised route networks:

- Recognise the importance of highways as part of the walking and riding network
- Include non-motorised networks as an integral part of a rural route hierarchy
- Replace difficult crossings and reconnect missing links
- Integrate highway, rights of way and transportation planning
- Develop a national code of practice for off-road routes

To prevent rural roads acquiring an urban character by giving more thoughtful attention to:

- Highway engineering; e.g. signs, lines, lighting, pollution, landscape
- Development on land adjacent to routes; e.g. adverts, fences, hedgerows
- Inappropriate or poorly designed development

How we can make things happen

Use better tools

- Use Placecheck as a community enablement/consultation tool
- Quiet Lanes and Greenway development
- Use the Government resources available

Share and use best practice

- Ensure that new policies and programmes are disseminated to a wide audience
- Insist on "joined-up thinking"
- Increase awareness of useful websites and technical papers

Allocate resources fairly

- Rebalance spending between motorised and non-motorised routes
- Include in the transport budget an allowance for "net-gain" for non-motorised routes
- Allocate a percentage of the highways budget for non-motorised routes

Work in partnership

- Encourage working with local organisations and the community
- Develop contacts through Local Access Forums
- Think outside the box: involve related user groups

Consider changing the law

- Give priority for pedestrians and riders over motorised traffic on certain routes

Change the mindsets

- Managing rural roads solely for the easier, quicker movement of motorised traffic, *versus* managing for non-motorised users, tourists, and tranquillity and resolving conflict with motorised users
- Wake up to the cumulative impact of incremental change.

Appendix I – Glossary

Bridleway	The off-road network for use by pedestrians, horse riders, and cyclists only
Footpath	Part of the off-road network for users on foot
Footway	Pedestrian right of way, next to the road
Greenway	Car-free off-road routes for shared use
Lane	Colloquial term for a narrow country road
Public footpath	Highway for use by walkers only
Public right of way	Those highways which are public footpaths, bridleways, roads used as public paths, and byways open to all traffic
Quiet Lane	A minor rural road which has been treated appropriately to enable shared use by walkers, cyclists, horse riders and motorised users
Rider	An equestrian or cyclist
Shared route	A road used by a variety of users including pedestrians, riders and drivers
Towpath	Route alongside canals which may or may not be a public right of way, usable by walkers and sometimes by cyclists on a permissive basis

Appendix II – Placecheck

General questions

1. **What do you like about the route?**
2. **What don't you like?**
3. **What needs to be improved?**
4. **Who needs to be involved?**
5. **What resources are available?**

Detailed questions

6. **How can we make this a more valuable route?**
   What is it used for?
   Who uses it?
   Who could use it?

7. **How can we make this a more special route?**
Maintenance
Is the route being well looked after?
Are there accumulations of litter, or agro-industrial waste?
Are fly-tips or dumped/burnt-out cars removed within one week?
Are verges being damaged by vehicle over-running?
Are eyesores being tackled?
Are stiles maintained?
Are surfaces safe and in good repair?
Is drainage attended to?

Magic
What can be done to make the place look special?
Are there valued buildings along the route?
What local styles or vegetation are in existence?
Should the history of the route be observed?
Can skylines, vistas or beauty spots be improved or created?

8. **How can we make it safer for people on foot, cycle or horse, or for farm and other animals?**
Are there opportunities to create safer conditions? e.g.
   Reducing the speed of traffic?
   Providing alternative routes?
   Creating greater separation between traffic and other users?
   Is it difficult or dangerous to cross the road?

9. **How can the route be made better-connected, or new routes created?**
How can the route be better linked in with the network of routes? e.g.:
   Safe routes to school?
   Safe routes to stations?
   Rural or Village Homezones?
What are the options to improve the value of this route by making changes to other routes?
For all types of users, is there a satisfactory range of:
   Leisure routes and circular walks – would new links help increase the range of walks (e.g. permissive footpaths)?
   Purposeful links to shops, stations, bus stops, etc?
   All-weather links – how can footpaths be made usable in the winter?

10. **How can we make the route more attractive?**
Reducing noise
Is noise a problem? Eroding tranquillity, interfering with wildlife?
Can traffic speeds be reduced?
Can a low-noise road surface be used?
Is there room to introduce noise barriers or bunds, without adversely affecting appearance?
Are there very local problems that can be tackled – e.g. noise from traffic accelerating away from a tight bend?
Can the volume of traffic be reduced?
Are there other sources of noise that need to be addressed?
Sensitive lighting
Is lighting effective, attractive and energy-efficient?
What does the lighting look like when viewed from local vistas?
Can shorter columns be used?
Can trees be planted near, to soften the impact on long-distance vistas?
Is light pollution (and hence energy use) minimised?
Does the lighting mark a community or a road?
Does privately owned lighting contribute or detract? Is the private lighting a source of light pollution?

Less clutter

Is there an accumulation of objects around the road that are making the place unattractive?

Can lines be removed?

Can the road be narrowed?

Can the surface be changed to a more natural-looking material – e.g. local stone dressing?

Do direction and other road signs reflect local traditions?

Can signs and street furniture be mounted on common posts or columns?

Can signs be eliminated? Do they serve a useful purpose or are they unnecessary?

Can the colour of columns be varied?

Improving the look of boundaries

Can the appearance be improved by introducing more natural or traditional ways of providing boundaries?

What are the local ways of providing boundaries in fields and woodland, and within villages?

Can fences be removed, reduced in size, or softened in appearance by growing plants in front or over them?

Can property entrances be made more attractive? (N.B. standard fences are ineffective at reducing traffic noise. They provide a minor barrier to criminals, and block surveillance)

Rivers, rills, streams and ditches

Are there any watercourses along the route?

Could they add to its attractiveness?

Are they easy to see from footpaths?

Are they being managed sensitively?

Buildings and gardens

Is new development reinforcing the character and attractiveness of the area?

Is the area covered by village design statements?

Are new buildings reflecting local styles and materials?

Does new development contribute to the local network of lanes and footpaths?

Are visibility splays required? Are they destroying existing hedging or walling?

Are small changes being well managed?

Are boundaries attractive?

Do gardens complement the surrounding countryside, or contrast?

11. How can the place better adapt to change?
Changing agriculture and forestry

What effect are changes in agriculture or farming having on the appearance of the local countryside?

Are fields increasing in size?

Are new trees being planted?

Are hedges or stone walls being maintained?

Are ponds being preserved or filled in?

Are areas of woodland being encroached upon?

How close does ploughing go to trees in fields?

Does anything need to be done?

How could the wider landscape be improved?

Appendix III – The PACE process

Use this process to plan and deliver Greenways and Quiet Lanes in your area. Further details can be found on the Countryside Agency Greenways and Quiet Lanes website at www.greenways.gov.uk

The four main components of the process are:

Plan:

- A vision for the Greenway Network
- Partner and stakeholder roles
- Policy Framework – national and local
- Demand assessment and data collection
- Community and user participation
- Objectives and targets for the network for non-motorised users including Greenways
- Develop a diagram indicating journey origins and destinations for pedestrian, cycling, equestrian, car and commercial traffic relating to the existing and proposed network and land-use pattern
- Assess the resource implications for the network

Activate:

Only when the policy and objectives of a network including Greenways are established can actions be formulated that lead to their fulfilment.

The Activate component develops a strategy to fulfil the policies and objectives and consists of the following elements:

- Public participation
- Environmental Quality (Assessment) and gathering of traffic data
- Management arrangements for users and infrastructure
- Resolution of legal issues
- Resolution of traffic issues
- Budget estimate
- Funding sources – capital and revenue
- Safety audit
- CDM – risk assessments

Check:

Prior to any enablement it is essential to check and verify the strategy. This should be done against the policy and objectives, against the criteria for route selection in the light of data gathering and through public participation in the strategy development and results.

Enable:

This is the implementation component and comprises:

- Land assembly and legal agreements
- Detailed design
- Costing
- Working drawings
- Statutory approval, planning, TROs
- Construction
- Accreditation validation and review
- Management and monitoring of the network in use with any actions proposed taken to the start of the process again for evaluation

The PACE process is cyclical. If the accreditation, network evaluation, target achievement or management indicates that the network is not maintaining quality or not meeting targets or objectives, then the process returns to the start and begins again.

Appendix IV – Useful contacts

There are many user groups, professional bodies and Government agencies providing good advice and ideas on how to change your local environment. The Internet in particular is a great place to find out more about how these organisations can help or provide advice.

User groups

British Horse Society

www.bhs.org.uk

Stoneleigh Deer Park
KENILWORTH CV8 2XZ
Tel: 0870 120 2244

The BHS works closely with Government bodies in order to improve Britain's bridleways.

Council for the Protection of Rural England

www.cpre.org.uk

128 Southwark Street
LONDON SE1 0SW
Tel: 020 7981 2800
Email: info@cpre.org.uk

The CPRE exists to promote the beauty, tranquillity and diversity of rural England by encouraging the sustainable use of land and other natural resources in town and country.

Cyclists' Touring Club

www.ctc.org.uk

Cotterell House
69 Meadrow
GODALMING GU7 3HS
Tel: 0870 873 0063
Email: cycling@ctc.org.uk

The CTC is a voluntary organisation, made up of 70,000 members, who organise cycling activities throughout the UK.

National Trust

www.nationaltrust.org.uk

36 Queen Anne's Gate
LONDON SW1H 9AS
Tel: 020 7222 9251
Email: enquiries@thenationaltrust.org.uk

The National Trust cares for over 248,000 hectares of countryside. The organisation encourages public access to its property wherever possible.

Ramblers' Association

www.ramblers.org.uk

2nd Floor, Camelford House
87-90 Albert Embankment
LONDON SE1 7TW
Tel: 020 7339 8500
Email: ramblers@london.ramblers.org.uk
(For Wales and Scotland replace London with the country's name)

The Ramblers' Association is a voluntary organisation dedicated to promoting the enjoyment and discoveries that walking in the countryside can bring.

Slower Speeds Initiative

www.slower-speeds.org.uk

PO Box 19
HEREFORD HR1 1XJ
Email: info@slower-speeds.org.uk

The Slower Speeds Initiative is a joint campaign between nine organisations to lower traffic speeds in order to benefit the community by reducing the harmful effects created by motor vehicles.

Sustrans

www.sustrans.org.uk

Head Office
35 King Street
BRISTOL BS1 4DZ
Tel: 0117 929 8893
Email: info@sustrans.org.uk

Sustrans is a charity that aims to encourage people to cycle and use public footpaths in order to reduce the adverse effects of motor vehicles.

Professional bodies

CSS

www.cssnet.org.uk

The CSS represents the directors of strategic planning, transportation, environment, waste management and economic development functions throughout the UK.

Institute of Public Rights of Way Officers

www.iprow.co.uk

PO Box 78
SKIPTON BD23 4UP
Tel: 0700 078 2318
Email: iprow@iprow.co.uk

IPROW is the professional body representing rights of way and countryside access officers in England and Wales who are responsible for the management of public access on the public rights of way network.

Institution of Civil Engineers

www.ice.org.uk

1 Great George Street
LONDON SW1P 3AA

The ICE is an independent engineering institution that represents over 80,000 qualified civil engineering professionals.

Institution of Highways and Transportation

www.iht.org

6 Endsleigh Street
LONDON WC1H 0DZ
Tel: 020 7387 2525
Email: iht@iht.org

The IHT is the foremost learned society in the UK concerned specifically with the design, construction, maintenance and operation of sustainable transport systems and infrastructure.

Landscape Institute

www.l-i.org.uk

6-8 Barnard Mews
LONDON SW11 1QU
Tel: 020 7350 5200
Email: mail@l-i.org.uk

The LI is the Chartered Institute in the UK for landscape architects. The field incorporates the design, management and science of enhancing and conserving the environment, and the planning, design and management of open spaces.

Royal Town Planning Institute

www.rtpi.org.uk

41 Botolph Lane
LONDON EC3R 8DL
Tel: 020 7636 9107
Email: online@rtpi.org.uk

The RTPI's core ideas are based around dealing with the needs and characteristics of places in ways that are sustainable, whilst also recognising the wide range of people who are involved in planning issues.

Urban Design Alliance

www.udal.org.uk

c/o The Urban Design Group
70 Cowcross Street
LONDON EC1M 6DG

The UDAL is a group of seven professional and specialist organisations working to create quality cities, towns and villages.

Government

Rights of Way Legislation,
Countryside Division 5
Department for Environment, Food and Rural Affairs
Zone 1/02, Temple Quay House
2 The Square
Temple Quay
BRISTOL BS1 6EB
Tel: 0117 372 8872
Email: www.rightsofway@defra.gsi.gov.uk

Government agencies

British Waterways

www.britishwaterways.co.uk

Willow Grange
WATFORD WD17 4QJ
Tel: 01923 226 422
Email: enquiries.hq@britishwaterways.co.uk

British Waterways actively maintain Britain's canals, rivers and towpaths for a wide range of business and leisure uses.

Countryside Agency

www.countryside.gov.uk

John Dower House
Crescent Place
CHELTENHAM GL50 3RA
Tel: 01242 521 381

The Countryside Agency works to conserve and enhance the countryside, to promote social equity and economic opportunity for people living in rural areas, and to help everyone to enjoy the countryside.

Forestry Commission

www.forestry.gov.uk

231 Corstorphine Road
EDINBURGH EH12 7AT
Tel: 0131 334 0303
Email: enquiries@forestry.gsi.gov.uk

The Forestry Commission of Great Britain is the Government department responsible for the protection and expansion of Britain's forests and woodland.

Working Party

John Thackray (Chair)	Institution of Civil Engineers/ Ramblers' Association
Wendy Otter	Countryside Agency
Robert Huxford	Institution of Civil Engineers
Paul Selby	Institution of Civil Engineers

With additional support from:

Janet Davis	Ramblers' Association
Janet Dickinson	Landscape Institute
Paul Hamblin	Council for the Protection of Rural England
Peter Hine	Royal Town Planning Institute/ Transport Planning Society
George Keeping	County Surveyors' Society
Jane Krause	Institute of Public Rights of Way
Emily Lauder	Royal Town Planning Institute
Paige Mitchell	Slower Speeds Initiative
Tony Russell	Cyclists' Touring Club
Julia Samson	Transport 2000
Tim Slade	Rights of Way, Isle of Wight Council
Mary Weston	British Horse Society
Don Matthew	Sustrans